FLORA OF TROPICAL EAST AFRICA

CHENOPODIACEAE

J. P. M. Brenan

Mostly annual or perennial herbs, sometimes shrubs, rarely small trees, often halophytic and with tendency to fleshiness. Leaves alternate, rarely opposite, simple, without stipules, entire or not, but characteristically never regularly serrate ; sometimes the leaves reduced to scales. Flowers small to minute, mostly green to grey, solitary and axillary, or more often variously clustered, usually regular, ⚥ or ♂ and ♀. Calyx (1–) 2–5-lobed, lobes more or less united below, imbricate or almost valvate, persistent after flowering, or variously modified in fruit. Petals absent. Stamens as many as or fewer than sepals, and opposite them. Ovary normally free and superior, 1-celled. Stigmas 2 (–5). Ovule solitary, campylotropous, on a long or short almost always basal funicle. Fruit usually utricular, indehiscent, rarely circumscissile, often included in and falling with the calyx. Embryo peripheral, curved, annular or spiral, surrounding the endosperm, which may be absent.

The presence or absence of endosperm and the position and shape of the embryo are characters very valuable for separating certain genera in this family. They are not employed in the following key, but are given in the generic descriptions.

Stems obviously leafy though the leaves sometimes very small, not appearing to be built up of green fleshy segments ; flowers not immersed in a fleshy spike ; plants mostly mealy, pubescent or glandular, sometimes glabrous :

Calyx not horizontally winged ; leaves often petio-late :

Plants (E. African species) obviously mealy, pubescent or glandular, at least on young parts ; flowers without bracteoles (except the ♀ in *Atriplex*) ; leaves various, normally comparatively broad, usually neither entire nor linear to linear-oblong :

Flowers ⚥ and ♀, all without bracteoles and with a calyx . . . . . . 1. **Chenopodium**

Flowers ♂ and ♀, ♂ without bracteoles and with a calyx, ♀ with 2 bracteoles but no calyx, the bracteoles enlarging and clasping the fruit . . . . . 2. **Atriplex**

Plants (E. African species) glabrous, rarely slightly and inconspicuously pubescent on very young-est parts only ; flowers all subtended by small scarious bracteoles ; leaves entire, linear to linear-oblong (in our species) ; flowers all sessile and axillary . . . . . 5. **Suaeda**

1

Calyx in fruit with a membranous horizontal wing
from the back of each of its segments ; leaves
sessile, in our species very small and entire ;
flowers hermaphrodite . . . . .   **6. Salsola**
Stems seemingly leafless, apparently built up of numerous superposed more or less tubular succulent green
segments, which ultimately shrivel and fall away
from the stem proper ; flowers immersed in a
fleshy spike ; plants glabrous :
Perennial with prostrate main stems (E. African
species), with flowers and fruits quite hidden by
the joints of the fertile spike, and remaining
attached to them when the spike breaks up into
separate joints at full maturity . . .   **3. Arthrocnemum**
Erect annual (E. African species), with groups of three
flowers or fruits clearly visible as flattened
triangles between the joints of the fertile spike,
which does not at full maturity break up into
separate joints ; fruits falling away from the
joints and leaving cavities . . . .   **4. Salicornia**

1. CHENOPODIUM

L., Gen. Pl., ed. 5, 103 (1754)

Mostly annual or perennial herbs, glabrous, pubescent, glandular or mealy
with vesicular hairs. Leaves alternate, mostly petiolate, normally broad.
Flowers mostly in cymose clusters (" glomerules ") variously arranged, ☿
and ♀ mixed, without bracteoles. Calyx of both sorts of flower normally
(3–) 4–5-lobed, unaltered or nearly so in fruit, or sometimes becoming fleshy.
Stamens 1–5. Fruits with membranous indehiscent pericarp. Seeds
" horizontal " (vertically compressed) or, less commonly, " vertical "
(horizontally compressed) ; testa normally thin, hard and brittle. Embryo
annular. Endosperm present.

Weeds of cultivated areas and waste lands around human habitations.

The species here have to be separated from one another with care. Differential
vegetative characters, though present, may be indefinite and difficult to portray. The
fruits and seeds, however, give for the majority of the species very precise and constant
characters. The markings on the testa of the seed, taxonomically very valuable, require
the low power of a compound microscope for them to be clearly seen. Special care

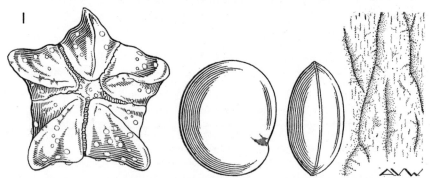

FIGS. 1 and 2 (on opposite page). *CHENOPODIUM*—Perianth containing fruit, seen from above, × 20 ;
seeds, front and side views, × 20 ; portion of surface of seed-testa, × 200. Species numbered as in text.
1, *C. album* ; 2, *C. opulifolium* ; 3, *C. murale* ; 4, *C. fasciculosum* ; 5, *C. ambrosioïdes* ; 6, *C. procerum* ;
7, *C. schraderianum* ; 8, *C. carinatum* ; 9, *C. pumilio*.

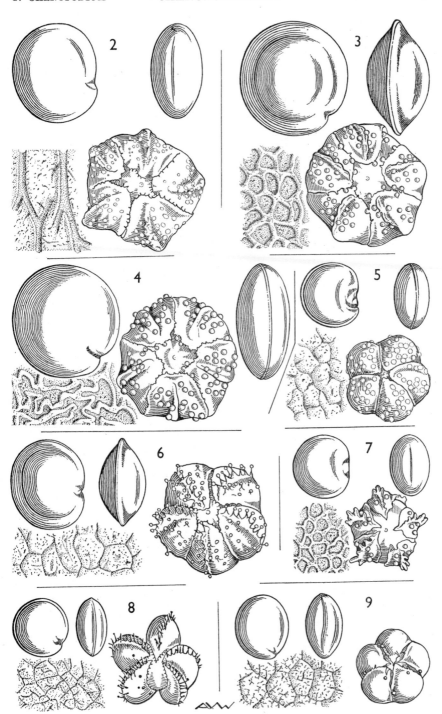

FIG. 2.   See caption of Fig. 1 on opposite page.

must be taken, before examining the testa, to remove the thin skin-like pericarp which closely covers it ; this can be done either by kneading some fruits between thumb and forefinger, or, if the pericarp is persistent, by scraping it away with needles, using a lens.   Until the user of this Flora is familiar with the species, it is better that he tries to name only plants bearing ripe seeds, which fortunately are lavishly produced.

In all the East African species, the flowers in each inflorescence are a mixture of hermaphrodite and female, the former usually occupying the terminal position in a cymule, the latter often opening later. In the following key and descriptions the stamen numbers must be taken to refer to the hermaphrodite flowers only.

Several of our species, especially *C. album*, *C. opulifolium*, *C. murale* and *C. ambrosioïdes*, also occur in Europe, and those seriously studying this difficult genus will do well to consult modern works dealing with it there. Hegi, Ill. Fl. Mittel-Eur. 3 (1910) is recommended for its illustrations ; Ascherson & Graebner, Syn. Mitteleur. Fl. 5 (1) (1913) for synonymy and an account of the wide ranges of variation of certain species ; while Clapham, Tutin & Warburg, Fl. Brit. Is. (1952) provides a concise and up-to-date account of the genus in Britain.

Plant more or less mealy, at least on young parts, with
    grey or whitish vesicular hairs ; other sorts of hair
    and also glands absent ; stamens (of hermaphro-
    dite flowers) always 5 ; seeds always black when
    ripe, 1 mm. or more in diameter :
  Seeds sharply keeled on margin, 1·2–1·5 mm. in
    diameter ; testa marked with very close minute
    rounded pits (Fig. 2/3) ; pericarp very difficult
    to detach from seed ; inflorescences always
    cymose and leafy .   .   .   .   .   .   3. *C. murale*

  Seeds bluntly keeled on margin ; testa not marked
    as above ; pericarp readily rubbed or scraped
    from seed :
    Seeds 1–1·5 mm. in diameter ; testa marked with
      radial furrows, and often also with minute
      roughnesses in between, never closely pitted :
      Leaves (except juvenile ones following the
        cotyledons) distinctly longer than broad,
        normally by at least 1½ times ; stems often
        more or less red ; branching commonly
        erect or suberect ; testa furrowed, other-
        wise almost smooth (Fig. 1) .   .   .   1. *C. album*

      Leaves (at least median and lower cauline)
        nearly or quite as broad as long, rather
        small, up to about 5·4 cm. long ; stems
        rarely red ; branching commonly diver-
        gent ; testa marked with radial furrows
        more closely than in *C. album*, also with
        minute roughnesses in between (Fig. 2/2) ;
        inflorescences normally very grey-mealy . 2. *C. opulifolium*

    Seeds 1·5–2 mm. in diameter ; testa furrowed or
      pitted :
      Leaves below widest point cuneate and normally
        entire, sometimes broadly cuneate ; teeth
        up to about 10 each side, usually fewer, not
        acuminate, usually directed upwards ;
        seeds not more than 1·85 mm. in diameter
        (usually less than 1·75 mm.) ; testa marked
        with radial furrows but not pitted (Fig. 1) ;
        calyces shed with fruit, sepals not becoming
        reflexed   .   .   .   .   .   .   1. *C. album*

Leaves below widest point rounded in outline to subtruncate or even subcordate and distinctly toothed ; teeth 7–60 each side, usually numerous, acuminate or acute, tending to be directed outwards ; seeds 1·5–2 mm. in diameter ; testa marked with very close minute sinuose and irregularly branched pits (Fig. 2/4) ; calyces often persisting on inflorescence after fruit is shed, their sepals reflexed and with thickened midribs . . . . . . 4. *C. fasciculosum*

Plant pubescent, and with yellow to amber glands, aromatic, without vesicular hairs ; stamens (of hermaphrodite flowers) 1–5 ; seeds black to red-brown when ripe, 0·5–1·25 mm. in diameter :

Inflorescence built up of distinct though sometimes small dichasial cymes in the axils of leaves or bracts, these cymes usually aggregated as though into a spike ; seeds black or nearly so when ripe ; stamens 1–2 ; lower and median leaves pinnately divided, at least their lower part ; sepals always keeled :

Seeds 0·7–0·8 mm. in diameter ; testa marked with very minute shallow contiguous rounded or angular pits (Fig. 2/7) ; glands between veins on lower surface of leaf, also those on outside of sepals, all sessile (use × 20 lens) ; leaves pinnately divided throughout each side usually to within 2–3 mm. of midrib . . 7. *C. schraderianum*

Seeds 0·9–1·1 mm. in diameter ; testa marked with slightly impressed sinuose lines and minor roughnesses (Fig. 2/6) ; glands between veins on lower surface of leaf, also many of those on outside of sepals, shortly but distinctly stalked (use × 20 lens) ; lower part of leaves pinnately divided, top part toothed but scarcely lobed . . . . . 6. *C. procerum*

Inflorescence built up of small sessile or subsessile clusters of flowers in the axils of leaves or bracts, flowers not in dichasial cymes ; seeds red-brown to blackish when ripe ; stamens 1–5 ; leaves and sepals various :

Sepals each having on its back outside a conspicuous wing-like keel broadening upwards ; leaves to 3 cm. long ; flowers all in leaf-axils ; stamen 1 ; seeds all " vertical " (see generic description), red-brown, 0·5–0·75 mm. in diameter . . . . 8. *C. carinatum*

Sepals rounded, not at all keeled on back :

Seeds in each cluster, some " vertical," others " horizontal " (see generic description), 0·5–1·25 mm. in diameter (in African specimens) ; stamens 4–5 ; ovary glandular above ; stigmas 3–4, long ; robust erect plant with paniculate inflorescence . . 5. *C. ambrosioïdes*

Seeds in each cluster all " vertical " (see generic
description), 0·5–0·75 mm. in diameter ;
stamen 1 ; ovary not glandular ; stigmas
2, short ; plant prostrate to ascending,
usually slender, with small leaves and
axillary flower-clusters not clearly panicu-
lately arranged   .   .   .   .   . 9. *C. pumilio*

1. **C. album** *L.*, Sp. Pl. 219 (1753) ; Bak. & C. B. Cl. in F.T.A. 6 (1) : 77
(1909), *pro parte, quoad specimen Forbes s.n. tantum* (= *C. album* f. *cymigerum*
(Koch) Schinz & Thell.) ; Hauman in F.C.B. 2 : 9 (1951). Type :
presumably from Europe, *Herb. Linnaeus* (LINN, lecto.!)

Herb usually 10–150 cm. high, annual, normally much branched but
sometimes lateral branches short and stem thus appearing subsimple, green
or red-tinged, especially on stem, more or less grey-mealy especially on
young parts. Leaves very variable, rhombic-ovate to lanceolate, mostly
distinctly longer than broad by normally at least 1½ times (but the juvenile
leaves following the cotyledons may be almost as broad as long), from about
1·2 × 0·3 and 2·3 × 1·3 cm. to 5·4 × 3·1, 6·5 × 3·8, 7 × 1·5, 7 × 3·7 and
8·2 × 5·5 cm., margins entire or more commonly with up to about 10 shallow
teeth each side, prominent basal lobes not commonly present, apex of leaf
acute, or particularly in the lower cauline leaves subacute to rounded ;
upper leaves and bracts progressively smaller. Inflorescence a usually
ample panicle of very numerous small or medium (2–6 mm. in diameter)
densely or laxly spicately or cymosely arranged dense rounded clusters
(" glomerules ") of minute grey to green flowers, latter 1–1·5 mm. in diameter.
Sepals 5, papillose on margins and outside, each with a prominent green keel
in upper part. Stamens 5. Pericarp somewhat persistent but easily rubbed
or scraped off seed. Seeds (Fig. 1, p. 2) black, shining, 1·2–1·6 (–1·85) mm.
in diameter, bluntly keeled, testa under microscope marked with faint radial
furrows further apart than in *C. opulifolium*, otherwise almost smooth.

KENYA. Kisumu-Londiani District : Londiani, July 1953, *C. van Someren* 194 ! &
C. van Someren in E.A.H. 10420 ! & Lumbwa, July 1953, *C. van Someren* 157 !
DISTR. **K5** ; widespread in the temperate and warm parts of the Northern Hemisphere ;
less common and possibly only an introduction further south, but found in S. America
and Australia ; in Africa (other than N.) recorded from the Belgian Congo, Portuguese
East Africa and S. Africa
HAB. A weed of cultivation, probably recently introduced

VARIATION. In Europe this species is excessively variable and polymorphic, but the
East African specimens appear typical. Other variants may well be introduced into
our area in the future.

NOTE. I have not given the extensive synonymy that there is for this species in Europe,
as it is not relevant here.

C. album is unquestionably very closely related to *C. opulifolium*, which is far
commoner in East Africa. *C. album* differs in the mature leaves being decidedly longer
than wide, often thinner in texture, and in the smoother surface of the seed when seen
under the microscope. In addition the inflorescence of *C. album* is normally greener and
not so densely glaucous-mealy as in *C. opulifolium*, and the stem is much more commonly
tinged with red. Some authors have stated that the branching of *C. opulifolium* is
divaricate, while that of *C. album* is stricter ; this is often a useful character, but I would
ask observers to test its constancy in East Africa.

2. **C. opulifolium** [*Schrad. ex*] *Koch & Ziz*, Cat. Pl. Palat. 6 (1814) ; DC.,
Fl. Fr. 6 : 372 (1815) ; Bak. & C. B. Cl. in F.T.A. 6 (1) : 78 (1909) ; Hauman
in F.C.B. 2 : 7 (1951). Type : Germany, Arheilgen, *Borkhausen* (location
unknown)

Herb up to 60–150 (–300) cm. high, annual or apparently sometimes a short-lived perennial which may become woody below, normally very much branched, green to almost white, rarely red-tinged, grey-mealy, sometimes densely so on young shoots, inflorescences and undersides of leaves. Leaves variable, mostly broadly and shortly rhombic-ovate, the median and lower almost as broad as long, from about (0·7 × 0·4 cm.–) 1 × 0·7–1 cm. to 3·7 × 2·8–3·7 cm. and 5·4 × 5·4 cm., in the lower third or about the middle normally with a short prominent often bilobed divergent lateral lobe each side, the margins above this lobe entire or with up to several teeth, apex of leaf rounded or acute ; in African specimens the lobes not uncommonly less prominent or even the leaves subentire ; upper leaves smaller and more acute. Inflorescence a panicle, usually ample, of very numerous small densely or laxly spicately or rarely cymosely arranged dense rounded clusters (" glomerules ") of minute grey to greenish flowers, latter 1–1·5 mm. in diameter. Sepals 5, papillose on margins and outside, each with a prominent green keel in upper part. Stamens 5. Pericarp somewhat persistent, but readily scraped off seed. Seeds (Fig. 2/2, p. 3) black, shining, 1·1–1·5 mm. in diameter, bluntly keeled, testa under microscope marked with radial furrows and minute irregular papillose roughening in between.

UGANDA. Karamoja District : Mt. Debasien, Jan. 1936, *Eggeling* 2840 ! ; Kigezi District : Kachwekano Farm, Feb. 1950, *Purseglove* 3246 ! ; Teso District : Serere, May 1932, *Chandler* 563 !
KENYA. Northern Frontier Province : Moyale, 18 July 1952, *Gillett* 13608 ! ; Naivasha District : Kedong, Mount Margaret Estate, June 1940, *Bally* 910 *in C.M.* 17088 ! ; Nairobi, 12 June 1930, *Napier* 296 *in C.M.* 1006 !
TANGANYIKA. Moshi District : Lyamungu, 24 Aug. 1932, *Greenway* 3137 ! ; Lushoto District : Makuyuni, *Koritschoner* 1060 ! ; Kondoa District : Sambala, 28 Mar. 1929, *Burtt* 2151 !
DISTR. U1–4 ; K1, 3–7 ; T1–7 ; Europe and the Mediterranean Region, eastwards to India and ? Mongolia, southwards through tropical Africa to Northern and Southern Rhodesia, Angola and ? S. Africa ; adventive in N. America
HAB. A weed of cultivation and settled areas ; 760–2100 m.

SYN. [*C. album* sensu Borkh. in Rhein. Mag. 1 : 472–3 (1793) ; Bak. & C. B. Cl. in F.T.A. 6 (1) : 78 (1909), *pro majore parte, excl. syn. C. murale, C. olukondae, et Kotschy* 8, *Pfund* 454, *non* L.]
    [*C. album* L. var. *viride* sensu Bak. & C. B. Cl. in F.T.A. 6 (1) : 78 (1909), *pro parte quoad Speke & Grant* 609, *non* (L.) Moq.]
    C. opulifolium [Schrad. ex] Koch & Ziz subsp. *ugandae* Aellen in F.R. 24 : 339 (1928). Type : Kenya, Nairobi, *Mearns* 283 (US, holo. !)
    C. mucronatum Thunb. var. *subintegrum* Aellen in F.R. 24 : 340 (1928), *pro parte ?, saltem quoad Holst* 8894, Tanganyika, Usambaras, Kwa Mshuza (Z, syn. !, K, isosyn. !), *Holst* 8924 (US, syn. !)
    [*C. opulifolium* [Schrad. ex] Koch & Ziz var. *olukondae* sensu Hauman in F.C.B. 2 : 8 (1951) *quoad descr. et spec. cit., saltem De Craene* 144, *non* (Murr) Hauman]
    C. opulifolium [Schrad. ex] Koch & Ziz subsp. *orientale* (" *orientalis* ") Murr in Mag. Bot. Lapok 1 : 339 (1902). Numerous syntypes cited, including three from our area : *Fischer, Stuhlmann, Stuhlmann* (B, syn. †)

VARIATION. The leaves are decidedly variable in the prominence of the lateral lobes and the amount of toothing. The frequent tendency in East Africa for this species to have inconspicuous lateral lobes and sometimes scarcely any teeth besides is quite inconstant even in our area, evidently of no systematic significance, and commonly no more than a state. I have seen closely similar specimens from Europe. Such plants have been called *C. mucronatum* var. *subintegrum* and *C. opulifolium* var. *olukondae*. Neither *C. mucronatum* Thunb. nor *C. olukondae* (Murr) Murr appear conspecific with *C. opulifolium* and neither of them apparently occurs in East Africa.

3. **C. murale** *L.*, Sp. Pl. 219 (1753) ; Bak. & C. B. Cl. in F.T.A. 6 (1) : 78 (1909) ; Aellen in F R. 24 : 343 (1928). Type : presumably from Europe, *Herb. Linnaeus* (LINN, lecto. !)

Annual up to 90 cm. high, upright or spreading, normally much branched, green, rarely red-tinged, mealy especially on young parts but rarely densely so. Leaves variable, commonly rhombic-ovate, rarely narrower, about 1·5–9 cm. long and 0·8–5 (–7) cm. wide, without any tendency for especially prominent basal lobes, but with several coarse irregular ascending usually sharp teeth (about 5–15 teeth on each margin, rarely fewer). Inflore cences leafy, composed of divaricately branched cymes up to 5 cm. long, terminal and from upper axils. Flowers greenish, minute, about 1–1·5 mm. in diameter. Sepals 5, papillose on margins and outside, each with a blunt raised green keel towards apex only. Stamens 5. Pericarp very difficult to detach from seed. Seeds (Fig. 2/3, p. 3) black, somewhat shining, 1·2–1·5 mm. in diameter, acutely keeled ; testa under microscope marked with very close minute rounded pits.

KENYA. Naivasha District : Gilgil R., Dec., *Scott Elliot* 6571 ! and Kedong, Mt. Margaret, June 1940, *Bally* 911a (mixed with 911) ! ; Nairobi, Marlborough Estate, Sept. 1939, *Bally* 114 *in C.M.* 11325 !
TANGANYIKA. Masai District : Ol Doinyo Sambu, Feb. 1926, *Haarer* B. 156 ! ; Moshi District : Ol Molog, 2 July 1945, *Greenway* 7520 ! ; Mpwapwa, 3 Mar. 1931, *Hornby* 381 !
DISTR. **K**3, 4 ; **T**2, 5 ; a cosmopolitan weed
HAB. A weed of crops, old cultivations and pasture land ; 1070–1950 m.

SYN. ? *C. longidjawense* A. Peter, F.D.O.-A. 2 : 204 & 22, t. 26, fig. 1 (1932 & 1938). Type : Tanganyika, Arusha District, Lengijawe, 1850–1880 m., *Peter* 42635 (B, holo. †)

VARIATION. *C. murale* is decidedly variable in size, habit and leaves, often doubtless due to habitat. Dr. Aellen has determined *Haarer* B. 156 (K !) as *C. murale* var. *spissidentatum* Murr in Mag. Bot. Lapok 2 : 11, t. 8, fig. 32 (1903) (*C. murale* f. *spissidentatum* (Murr) Aell. in Verh. Naturf. Gesellsch. Basel 41 : 87 (1930). This represents a form with the teeth on the leaves rather more numerous, closer and more prominent than usual, but the difference is at best but slight and can readily be matched in Europe. For the present it is preferably considered, with other comparable extremes known elsewhere but lacking any geographical correlation, which will very probably turn up in our area, as falling within the normal range of variation of *C. murale*.

In spite of its Latin name, in Africa as in England, it is no use searching for *C. murale* on walls.

C. longidjawense A. Peter is probably *C. murale*, but the type being no longer extant, there is an element of doubt. The hairs on some of the anthers, noted by Peter, are probably fungal.

4. C. fasciculosum *Aellen* in F.R. 24 : 344 (1928). Types : Kenya, Naivasha District, S. side of L. Naivasha, *Mearns* 672 (US, syn. !), 707 (US, syn., BM, isosyn. !)

Annual herb up to 2 m. high, upright, rarely spreading, normally much branched, green or slightly red-tinged, mealy especially on young parts but soon sparingly so or glabrescent. Leaves (in outline) ovate or deltoid-ovate, mostly 2·5–15 cm. long and 1·5–10 (–12) cm. wide, with more or less numerous irregular often coarse sharp teeth (7–60 or more on each margin). In-florescences composed of divaricately branched cymes up to 5 cm. long from upper leaf-axils and also sometimes terminal. Flowers greenish or red-tinged, minute, about 0·8–1·5 mm. in diameter. Sepals 5, papillose on margins and particularly towards base outside, each with a rather narrow blunt raised green keel towards apex only. Stamens 5. Pericarp easily scraped off seed. Seeds (Fig. 2/4, p. 3) black, rather shining, 1·5–2 mm. in diameter, bluntly keeled ; testa under microscope marked with very close minute irregularly but not radially elongate pits which are sinuose and irregularly branched. Fig. 3.

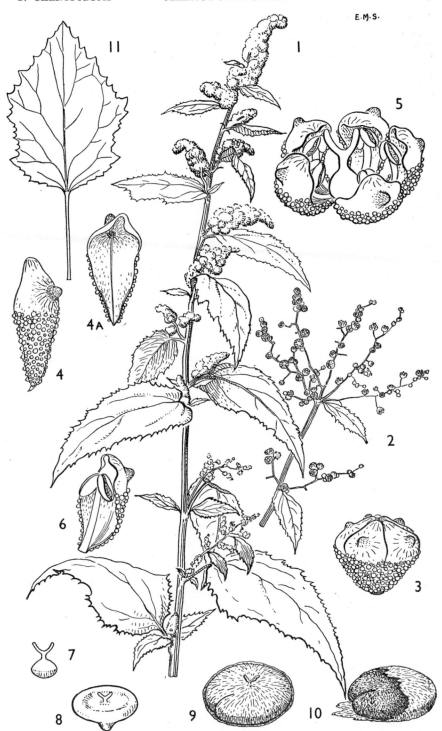

FIG. 3. *CHENOPODIUM FASCICULOSUM* var. *FASCICULOSUM*—1, part of flowering plant, 2/3 natural size ; 2, part of fruiting plant, × 2 ; 3, flower-bud, × 24 ; 4, sepal, seen from side, × 24 ; 4A, sepal, seen from inside, × 24 ; 5, open flower, × 24 ; 6, stamen opposite sepal, × 24 ; 7, young ovary, × 18 ; 8, older ovary, × 18 ; 9, fruit, × 18 ; 10, fruit with portion of the skin-like pericarp removed so as to expose seed, × 18. *CHENOPODIUM FASCICULOSUM* var. *MURALIFORME*—11, leaf, 2/3 natural size. 1, 2, 7, 8, from *Geilinger* 4001 ; 3–6, 9, 10, from *Wallace* 1103 ; 11, from *Bogdan* 1933.

var. fasciculosum

Leaves acuminate at apex. Teeth (on well-developed leaves) more numerous than in var. *muraliforme*, generally in all (16–) 20–60 or more on each margin, mostly acuminate ; little teeth numerous.

KENYA.   Northern Frontier Province : Mt. Kulal, 10 Oct. 1947, *Bally* 5606 ! ; Kiambu District : Kabete, 13 Apr. 1945, *Nattrass in Bally* 4287 !
TANGANYIKA.   Masai District : Ngorongoro, *Bally* 2419 ! ; Arusha District : Engare Olmotoni, June 1926, *Haarer* 246 ! ; Moshi District : Lyamungu, 28 Oct. 1943, *Wallace* 1103 !
DISTR.   K1, 3, 4 ; T2 ; Ethiopia

SYN.   [*C. murale* sensu Bak. & C. B. Cl. in F.T.A. 6 (1) : 79 (1909), *pro parte, quoad specimina Johnston, Quartin Dillon & Petit* 315, *Schimper* 357, *non* L.]

var. **muraliforme** Aellen in F.R. 24 : 344 (1928). Type : Kenya, Mt. Kenya, 2550 m., westward to the Kasorongai R., 1950 m., *Mearns* 1814 (US, holo. !)

Leaves acute at apex. Teeth (on well-developed leaves) comparatively few, generally 7–20 (–25) on each margin, acute, mostly coarse, little teeth few.

KENYA.   Nakuru District : Elburgon, Sept. 1933, *Napier* 2843 *in C.M.* 5252 ! ; Kiambu District : Limuru, Dec. 1927, *Lyne Watt* 1189 ! ; Kericho District : Lumbwa, July 1934, *Napier* 3406 *in C.M.* 6640 !
DISTR.   K3, 4, 5 ; not known elsewhere. May occur in K6 on the Mau ; confirmation wanted
HAB. (of species as whole).   Insufficiently known ; recorded from upland evergreen forest, by roadsides, and as a weed in cultivated areas and waste places ; 1310–2440 m.

This very interesting species, known outside our area only from Ethiopia, is related to *C. murale*, whose leaves are often more cuneate at base than in *C. fasciculosum* and with different toothing. The seeds of *C. murale* are, however, very different, smaller, sharply keeled, with a persistent pericarp and rounded not elongate-branched pits on the surface of the testa. This is a good example of a frequent occurrence among goosefeet : where constant seed-differences reinforce vegetative differences which though clear are hard to convey in words.
Aellen in his original description of *C. fasciculosum* gives the range in size of seed as 1·25–2 mm. ; I have not seen ripe seeds less than 1·5 mm.

5. **C. ambrosioïdes** *L.*, Sp. Pl. 219 (1753) ; Bak. & C. B. Cl. in F.T.A. 6 (1) : 79 (1909) ; Ulbr. in E. & P. Pf. ed. 2, 16c : 491, Fig. 183 K–Q (1934) ; Aellen & Just in Amer. Midl. Nat. 30 : 50–51 (1953) ; Hauman in F.C.B. 2 : 2 (1951).   Type : Spain, *Herb. Linnaeus* (LINN, lecto. !)

Herb up to 120 cm. high, usually annual, rarely a short-lived perennial, polymorphic (principally in America), upright, much branched, green (? occasionally red-tinged), variably pubescent or hairy especially on stem, also with numerous yellowish sessile glands particularly on lower side of leaves, strongly aromatic. Leaves variable, lanceolate in outline to more rarely elliptic or obovate, mostly 1·5–10 cm. long and 0·4–4·3 cm. wide, entire to laciniate or pinnatifid ; upper leaves and bracts smaller and narrower. Inflorescence an ample much-branched panicle with small sessile flower-clusters arranged spicately along the ultimate branches. Flowers greenish, 0·5–1·5 mm. in diameter. Sepals 3–5 (♀ flowers) 4–5 (⚥ flowers), pubescent to glabrous, glandular, variably connate, smooth or very rarely keeled. Stamens 4–5. Pericarp easily removed. Seeds deep red-brown to blackish or shining, 0·5–1·25 (–1·5) mm. in diameter, bluntly keeled, testa under microscope almost smooth or shallowly and irregularly pitted, usually marked also with sinuose lines.

forma ambrosioïdes

Shortly and inconspicuously pubescent or puberulous, sometimes with some longer hairs on stem. Leaves mostly lanceolate, lower with more or less coarse irregular ascending teeth, upper becoming entire. Flower-clusters all or mostly bracteate. Sepals not at all keeled. Seeds (Fig. 2/5, p. 3) deep red-brown, about 0·5–0·8 mm. in diameter, testa under microscope almost smooth and with faintly impressed sinuose lines.

Uganda. Masaka District : Koki, Buyamba, Aug. 1945, *Purseglove*, 1787 ! ; Mengo District : Kampala, 26 May 1933, *A. S. Thomas* 1219 !

Kenya. Nairobi, July 1943, *Nattrass 254 in C.M.* 15650 !

Tanganyika. Shinyanga, *Koritschoner* 2006 ! ; Tabora District : Nyembe, 6 June 1913, *Braun* 5417 !

Distr. U4 ; K4 ; T1, 4, 6 ; throughout the tropical and subtropical regions of the world ; especially polymorphic in America

Hab. A weed near habitations, sometimes on permanent ways. Certainly not native ; altitude range incompletely known, but recorded at about 1220 m.

Our plants are all, according to the nomenclature of Aellen in F.R. 26 : 33–34 (1929) and of Aellen & Just in Amer. Midl. Nat. 30 : 51–52 (1943), *C. ambrosioides* subsp. *eu-ambrosioides* Aellen var. *typicum* (Speg.) Aellen f. *genuinum* (Willk.) Aellen. Hauman in F.C.B. 2 : 3 (1951) reports the same for the Belgian Congo. The uniformity in Africa is in strong and suggestive contrast with the great variation in America. Since all the ranks of the hierarchy given by Aellen include the type of the species, for each and all the epithet must be *ambrosioides* unaltered.

The conception of *C. ambrosioides* adopted here follows the papers of Aellen and of Aellen & Just already mentioned, and is not that of Woroschilov's revision of the species of *Chenopodium* sect. *Ambrina* (Spach) Hook. f., in Journ. Bot. U.R.S.S. 27 : 33–47 (1942).

Besides in the points mentioned in the description above *C. ambrosioides* differs from *all* the other East African species in the ovary being glandular above not smooth, and in the stigmas being mostly 3–4, not constantly 2.

6. C. procerum [*Hochst. ex*] *Moq.* in DC., Prodr. 13 (2) : 75 (1849) ; Hauman in F.C.B. 2 : 6 (1951). Types : Ethiopia, Tigré, near Adowa, *Schimper* 882, 1506 (G, syn., K, isosyn. !)

Herb, probably annual, up to 2 m. or more high, upright, with few or many branches, green or often strongly red-tinged, glandular-pubescent all over, usually shortly so, strongly aromatic. Lower and median leaves elliptic or ovate-elliptic in outline, mostly 2·5–14 cm. long and 1·5–7 (–9) cm. wide, acute at apex, pinnately divided each side into 3–5 more or less sharply-toothed lobes, the lower lobes extending to near midrib, the upper much shallower, and the top part of the leaf normally toothed but scarcely lobed, glands between veins on lower surface shortly but distinctly stalked (use × 20 lens) ; upper leaves smaller, narrower and less divided, uppermost ones often oblong and obscurely sinuate-dentate. Inflorescences composed of dichotomously branched axillary cymes which are usually aggregated into more or less leafy or nude continuous cylindrical inflorescences 1·5–6 cm. wide and up to 60 cm. or more long. Flowers greenish or red-tinged, minute, 0·5–1·5 mm. in diameter. Sepals 5, each with a green glandular keel towards apex, glandular outside, many glands distinctly stalked (× 20 lens necessary). Stamens 1–2. Pericarp easily scraped off seed. Seeds (Fig. 2/6, p. 3) black or nearly so, glossy, 0·9–1·1 mm. in diameter, with a rather prominent but blunt keel ; testa under microscope very slightly rough with slightly impressed irregular sinuose lines and other minor roughnesses.

Uganda. Kigezi District : Kachwekano Farm, Jan. 1950, *Purseglove* 3227 ! ; Elgon, Jan. 1918, *Dummer* 3630 ! ; Masaka District : Kabula, Sept. 1945, *Purseglove* 1823 !

Kenya. Uasin Gishu District : Kipkarren, Sept. 1931, *Brodhurst-Hill* 329 *in C.M.* 3628 ! ; Nakuru District : Rongai, 23 July 1951, *Bogdan* 3162 !

Tanganyika. Bukoba District : Bugufi, Jan. 1936, *Chambers* K25 ! ; Mbulu District : Oldeani, 20 June 1935, *R. M. Davies* 1065 ! ; Lushoto District : Makuyuni, *Koritschoner* 1335 !

Distr. U2–4 ; K3, 6 ; T1–3 ; the A.-E. Sudan and Ethiopia southwards through eastern Africa to Nyasaland

Hab. Upland grasslands, and a local weed of cultivated areas and waste places ; 1340–2440 m.

Syn. *C. suberifolium* Murr in Bull. Herb. Boiss., sér. 2, 4 : 990, t. 6, fig. 3a, b. (1904) ; Bak. & C. B. Cl. in F.T.A. 6 (1) : 80 (1909) ; Aellen in F.R. 24 : 347 (1928) ;

F.D.O.-A. 2 (1) : 204 (1932). Type : Tanganyika : Lushoto District, Kwa Mshusa, *Holst* 8926 (Z, holo., K, iso. !)
C. botrys L. var. *procerum* [" *procera* "] (Hochst. ex Moq.) C. B. Cl. in F.T.A. 6 (1) 80 (1909) ; F.D.O.-A. 2 (1) : 204 (1932)
[*C. botrys* sensu Bak. & C. B. Cl. in F.T.A. 6 (1) : 80 (1909), *pro parte quoad Scott-Elliot* 6624, *Buchanan* 845 et *Whyte* (Nyasaland), *non* L.]

C. procerum has been confused with *C. schraderianum*, with which it has much in common, but is certainly distinct in a number of ways : (a) tendency to greater stature and more branches, (b) leaves mostly acute not obtuse, with different lobing, (c) glands on lower surface of leaf between veins, also on outside of sepals, mostly stalked not sessile, (d) seeds slightly but constantly larger, and with somewhat more prominent keel, (e) testa marking different—compare descriptions and Fig. 2.
Hauman's statement in F.C.B. 2 : 7 (1951) that *C. procerum* differs from *C. schraderianum* solely in leaf-shape is inadequate.

7. C. schraderianum *Schult.*, Syst. Veg. 6 : 260 (1820) ; Bak. & C. B. Cl. in F.T.A. 6 (1) : 80 (1909) ; Brenan in K.B. 1950 : 131 (1950) ; Hauman in F.C.B. 2 : 4 (1951). Type : a plant cultivated by Schrader, of unknown origin

Annual up to 1–1·3 m. high, upright ; main stem simple or with few rarely many lateral branches, especially near base ; plant green, sometimes red-tinged, shortly glandular and pubescent all over, strongly aromatic. Lower and median leaves elliptic to oblong in outline, mostly 1–5 (–8) cm. long and 0·5–3 (–5) cm. wide, mostly obtuse at apex, pinnately divided throughout each side into 3–5 narrow blunt lobes which are entire or with a few blunt teeth and all extend usually to within 2–3 mm. of midrib, glands between veins on lower surface of leaf all sessile (use × 20 lens), not accompanied by hairs ; upper leaves progressively smaller and less divided. Inflorescence as in *C. procerum* (p. 11) but often tending to be smaller. Flowers greenish or red-tinged, minute, 0·5–1 mm. in diameter. Sepals 5, each with a prominent toothed keel outside from near apex to near base, glandular outside, glands all sessile (use × 20 lens). Stamens 1–2. Pericarp easily rubbed or scraped off seed. Seeds (Fig. 2/7, p. 3) black or nearly so, somewhat glossy, 0·7–0·8 mm. in diameter, bluntly and not prominently keeled ; testa under microscope with very minute shallow contiguous rounded or angular pits.

UGANDA. West Nile District : Attiak, War, Apr. 1940, *Eggeling* 3911 ! ; Kigezi District : Kachwekano Farm, Sept. 1949, *Purseglove* 3108 ! ; Masaka District : Kyotera, Nov. 1945, *Purseglove* 1865 !
KENYA. North Nyeri District : Nyeri, 19 Dec. 1921, *Fries* 138 ! ; Nairobi, Kirichwa ndogo valley, Jan. 1940, *Bally* 749 ! ; Nakuru District : Londiani, Dec. 1905, *G. S. Baker* 348 !
TANGANYIKA. Bukoba District : Nyaishozi, Dec. 1931, *Haarer* 2443 ! ; Mbulu District : Mbulumbul, 25 June 1945, *Greenway* 7445 ! ; Rungwe District : Kyimbila, 23 Sept. 1910, *Stolz* 292 !
DISTR. U1, 2, 4 ; K3–7 ; T1–3, 7 ; from the A.-E. Sudan, Ethiopia and Somaliland southwards through eastern Africa to the Cape and Angola ; rarely adventive in Europe ; recorded from other parts of the Old and New Worlds, but apparently in error
HAB. A locally common weed of cultivated areas and waste places ; 1220–2290 m.

SYN. *C. foetidum* Schrad. in Magaz. Ges. Naturf. Freunde Berlin 2 : 79 (1808) ; Aschers. & Graebn., Syn. Mitteleur. Fl. 5 (1) : 24 (1913) ; Aellen in F.R. 24 : 345 (1928) ; Aellen & Just in Amer. Midl. Nat. 30 : 56 (1943), *non* Lam., Fl. Fr. 3 : 244 (1778)

For the differences between this species and *C. procerum* [Hochst. ex] Moq. see under the latter (above).
A number of varieties and forms have been recognized under *C. schraderianum*, notably by Murr in Bull. Herb. Boiss., sér. 2, 4 : 990–991 (1904), Aschers. & Graebn., Syn. Mitteleur. Fl. 5 (1) : 24 (1913), and Aellen in F.R. 24 : 345–7 (1928) ; these have been summarized by Hauman in F.C.B. 2 : 6 (1951). None has so far been recorded for East Africa, although doubtless they occur. Hauman's view that these are nothing more than states due to environment, or merely different stages in growth and development, is willingly followed here.

8. **C. carinatum** *R.Br.*, Prodr. Fl. Nov. Holl. 1 : 407 (1810) ; Aellen in Verh. Naturf. Gesellsch. Basel 44 (1) : 312 (1933) ; Ulbr. in E. & P. Pf., ed. 2, 16c : 494 (1934). Type : Australia, New South Wales, Hawkesbury R., *R. Brown* 3034 (BM, holo. !)

Annual herb similar to *C. pumilio*, prostrate to erect, normally branched near base into simple to much branched stems up to 60 cm. long, green rarely red-tinged, pubescent and glandular, aromatic. Leaves ovate to elliptic rarely narrow-elliptic in outline, small, mostly 0·3–3 cm. long and 0·25–2 cm. wide, sometimes as wide as long, with 2–4 (–6) usually coarse sometimes obscure entire or denticulate teeth or lobes on each margin, rarely entire or almost so, glands between veins on lower side of leaves sessile to sub-sessile not accompanied by hairs unless on veins. Flowers greenish, minute, about 0·4–0·75 mm. in diameter, sessile or subsessile in small rounded axillary clusters at most of the nodes. Sepals normally 5, pubescent and glandular, each with a conspicuous wing-like keel broadening upwards. Stamen 1. Seed (Fig. 2/8, p. 3) as in *C. pumilio*.

KENYA. Trans-Nzoia District : 32 km. SE. of Kitale, 21 June 1948, *Bogdan* 1738a! ; Nairobi : Karura Forest, 14 Sept. 1951, *Bogdan* 3263! ; Nairobi–Kiambu road, 3 Dec. 1950, *Verdcourt* 391!
DISTR. K3, 4 ; native of Australia, doubtfully so in New Zealand and New Caledonia ; introduced, often through wool, into other parts of the world including Britain.
HAB. A weed by roadsides and among cultivated maize ; also recorded from bare patches of rock in *Aristida–Harpachne* grassland ; altitude-range uncertain, recorded from 1620–1680 m.

Aellen in Verh. Naturf. Gesellsch. Basel 44 (1) : 312–3 (1933) splits *C. carinatum* into two varieties—var. *carinatum* (which he called var. *holopterum* (Thell.) Aell.) and var. *melanocarpum* (Black) Aellen, differing in their perianth-segments. The var. *melanocarpum* is treated as a distinct species, *C. melanocarpum* (Black) Black, by J. M. Black in Fl. S. Austral., ed. 2, 2 : 289 (1948). *Verdcourt* 550! (Nairobi) and *Bogdan* 3263! are certainly var. *carinatum* ; but other specimens do not show fully mature fruits and may or may not be the same variety.

9. **C. pumilio** *R.Br.* Prodr. Fl. Nov. Holl. 1 : 407 (1810) ; Benth., Fl. Austral. 5 : 163 (1870) ; Aellen in Verh. Naturf. Gesellsch. Basel 44 (1) : 314 (1933) ; Ulbr. in E. & P. Pf., ed. 2, 16c : 494 (1934) ; Black, Fl. S. Austral., ed. 2, 2 : 289 (1948). Type : Australia, *R. Brown* (BM, holo. !, K, iso. !)

Annual herb, prostrate or ascending, with numerous usually slender simple to much branched stems 2–45 cm. long radiating from rootstock, green rarely red-tinged, pubescent and glandular, aromatic. Leaves elliptic to lanceolate in outline, small, mostly 0·3–2·7 cm. long and 0·2–1·3 cm. wide, with 2–4 (–5) usually coarse sometimes obscure teeth or lobes on each margin, rarely (on depauperate plants) entire or almost so, glands between veins on lower side of leaves sessile to subsessile, not or only sometimes accompanied by glandular hairs except on veins. Flowers greenish, minute, about 0·4–0·75 mm. in diameter, sessile or subsessile in small rounded axillary clusters at most of the nodes. Sepals 4–5, sessile-glandular especially below, and more or less pubescent above, not at all keeled, but in fruit becoming rounded-convex on back outside and often whitish or pallid. Stamen 1. Pericarp easily scraped off seed. Seeds (Fig. 2/9, p. 3) all " vertical " (laterally compressed), deep red-brown, shining, 0·5–0·75 mm. in diameter, bluntly or sharply keeled ; testa under microscope almost smooth.

KENYA. Uasin Gishu District : 6 km. E. of Eldoret, 28 Aug. 1948, *Bogdan* 1885! ; Nairobi, Kabete, 28 May 1947, *Bogdan* 558! ; Kericho District : Kericho, 19 July 1933, *Napier* 2681 *in C.M.* 5096!
DISTR. K3–6 ; native of Australia, also in New Zealand and New Caledonia ; frequently introduced with wool into other parts of the world, e.g. the United States and Europe, including Britain

Hab. A weed of cultivated areas and by railways, in some places very common ; most probably introduced with agricultural seeds ; 1650–2530 m.

Syn. [*C. carinatum* sensu auct. mult., pro parte, non R.Br.]

The East African specimens are typical *C. pumilio* and not f. *glandulosum* (Moq.) Aellen in Verh. Naturf. Gesellsch. Basel 44 (1) : 315 (1933).

J. M. Black, Fl. S. Austral., ed. 2, 2 : 289 (1948), separates the common form of *C. pumilio* with deeply lobed leaves 5–20 mm. long as var. *oblongifolium* Black, the type of *C. pumilio* having very small entire or very shallowly lobed leaves 3–8 mm. long. The var. *oblongifolium*, Miss C. M. Eardley kindly informs me, has not been provided with a valid Latin description, but whether the type of *C. pumilio* represents a constant variant or a mere state has still to be decided ; for the present therefore var. *oblongifolium* is not adopted here, though later research may show it to be valid.

2. **ATRIPLEX**
L., Gen. Pl., ed. 5, 472 (1754)

Annual or perennial herbs or shrubs, usually more or less mealy with vesicular hairs. Leaves alternate, rarely opposite, mostly petiolate, normally broad. Flowers in clusters, which are axillary or aggregated into terminal panicles or spikes, monoecious or dioecious. ♂ flowers without bracteoles, with a 3–5-lobed calyx and 3–5 stamens ; ♀ flowers with 2 relatively large bracteoles, without calyx or staminodes. Fruits enclosed by the persistent accrescent and often modified bracteoles, with a membranous indehiscent pericarp. Seeds almost always " vertical " (i.e. laterally compressed), testa commonly thin and hard. Embryo annular with inferior radicle pointing upwards or upwards and outwards. Endosperm present.

Brackish or saline soil indicators.

Leaves more or less cordate or with obtuse auricles at
    base ; inflorescence an irregular terminal panicle ;
    bracteoles (in E. African specimens) muricate or
    tuberculate in centre, basal part thickened, obconical
    or campanulate .   .   .   .   2. *A. farinosa*
Leaves cuneate at base :
    Fruiting bracteoles when ripe with fleshy normally
        cherry-red lower part, smooth or nearly so on
        back, dried ones brown, veiny and flattened ;
        stems prostrate ; flowers in axillary clusters ;
        leaves up to about 1 cm. wide .   .   .   .   3. *A. semibaccata*
    Fruiting bracteoles when ripe neither fleshy nor red,
        ripe ones greenish to whitish when dry ; stems
        ascending, erect or bushy ; leaves often more than
        1 cm. wide :
        Leaves entire or with a single lateral lobe on each
            margin ; fruiting bracteoles (in E. African
            specimens) conspicuously muricate or tubercu-
            late in centre ; shrub or woody herb   .   .   1. *A. halimus*
        Leaves normally with 1–4 coarse teeth on each
            margin ; fruiting bracteoles smooth in centre ;
            herb, usually annual .   .   .   .   .   4. *A. muelleri*

1. **A. halimus** *L.*, Sp. Pl., ed. 1, 1052 (1753) ; Aschers. & Graebn., Syn. Mitteleur. Fl. 5 (1) : 150 (1913) ; Aellen in E.J. 70 : 11 (1939). Type : without locality, *Herb. Linnaeus* (LINN, lecto. !)

Shrub or woody herb, much branched, 0·5–3 m. high, densely mealy all over so that the whole plant is whitish to pale grey-green. Leaves ovate to oblong or elliptic, mostly 1–4 cm. long, 0·4–2·5 (–3) cm. wide, rounded to acute at apex, broadly or narrowly cuneate at base, entire or sometimes with a projecting lateral lobe on each side in lower part. Inflorescence a terminal panicle with spiciform mostly leafless ultimate branchlets. Fruiting bracteoles sessile, their basal connate part absent or very short indeed (to 1·5 mm.) ; free apical part of each bracteole reniform to broadly deltoid-ovate, as broad as or more commonly broader than long, 2·5–6 mm. long, 3–8 mm. wide, acute to rounded at apex, cordate or sometimes truncate or very broadly cuneate at base.

var. **granulata** *Chevall.* in Bull. Herb. Boiss., sér. 2, 5 : 444 (1905). Type : Algeria, Biskra, *L. Chevallier* 614 (? G, ? lecto., K, ? isolecto. !)

Fruiting bracteoles all or nearly all conspicuously muricate or tuberculate in the centre, broader than long, margins more or less denticulate or almost lobed.

TANGANYIKA. Mbulu District : Mbugwe, 18 June 1941, *R. M. Hornby* 2137 ! ; Singida District : Mkalama, near Sekenke Gold Mine, Oct. 1935, *Burtt* 5245 !
DISTR. **T**2, 5 ; Algeria ; to be sought for elsewhere
HAB. Flood-plain, pan and valley grasslands ; recorded as growing on seasonally flooded saline soils with short scanty grass, and on overgrazed areas of grey-white soil ; about 910–1220 m. ?

DISTR. (of species as a whole). Mediterranean Region from France and Spain to Syria and Transjordan ; North Africa ; Tanganyika ; apparently in Madagascar

The material from our area is not representative, and its final status must be settled by further collecting. It does not seem specifically separable from *A. halimus*, which in its commonest form has non-tuberculate, entire or somewhat denticulate fruiting bracteoles. Occasional bracteoles lacking tubercles are present on the Tanganyika plants. The inflorescence on Linnaeus' type bears apparently only galls in very young state.
The South African plants called *A. halimus* in the Flora Capensis are mostly *A. vestita* (Thunb.) Aellen, which is very near but with much larger fruiting bracteoles (see Aellen in E.J. 70 : 391 (1940)).

2. **A. farinosa** *Forsk.*, Fl. Aegypt.-Arab. cxxiii (1775) ; Boiss., Fl. Or. 4 : 917 (1875) ; Bak. & C. B. Cl., in F.T.A. 6 (1) : 82 (1909) ; Blatter, Fl. Aden (in Rec. Bot. Surv. India 7 :) 307 (1915) ; Schwartz, Fl. Trop. Arab. (in Mitteil. Inst. Allg. Bot. Hamburg 10 :) 35 (1939) ; Aellen in E.J. 70 : 8 (1939). Type : " Foliis cordato-ovalibus, retusis, crassiusculis, farinosis," from sandy plain by sea, Yemen, Arabia (Forsk., *l.c.*) ; no specimen in Forskål's herbarium (see C. Christensen in Dansk. Bot. Arkiv 4 (3) : 37 (1922)).

Shrub or woody herb 1–2 m. high, densely mealy all over so that the whole plant is whitish to pale grey-green. Leaves mostly ovate-oblong, 2–6 cm. long, 1–3 (–3·5) cm. wide, rounded at apex, more or less cordate or with obtuse auricles at base, margins entire or almost so ; petiole 1–4 mm. long ; upper leaves smaller and narrower, sometimes subacute at apex and ends of auricles. Inflorescence an irregular terminal panicle, the ultimate branchlets bearing dense sessile flower-clusters spiciformly arranged. Connate basal part of fruiting bracteoles thickened, 1·5–4 mm. long, 1·5–3 mm. wide, obconical or campanulate, the free apical part of each bracteole rounded to ovate or triangular, 3–5 mm. long and wide, obtuse to subacute at apex and with entire margins.

var. **keniensis** *Brenan* in K.B. 1953 : 432 (1953). Type : Kenya, Kilifi, 5 Nov. 1945, *Jeffery* K380 (K, holo. !)

Fruiting bracteoles conspicuously muricate or tuberculate in the centre, at the base of their free part. Fig. 4 (p. 16).

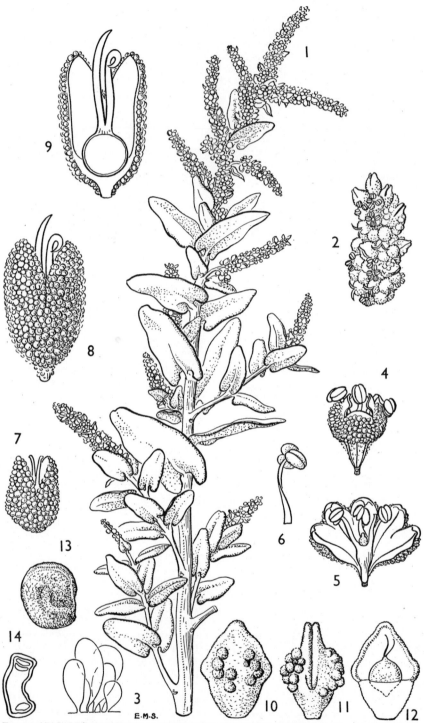

FIG. 4. *ATRIPLEX FARINOSA* var. *KENIENSIS*—1, part of plant, 2/3 natural size ; 2, part of inflorescence showing ♂ and ♀ flowers, also the enlarging bracteoles of the ♀ flowers, × 3 ; 3, bladder-like hairs from outside of sepal, × 100 ; 4, ♂ flower, × 8 ; 5, ♂ flower opened, showing stamens (one anther removed) and rudimentary ovary, × 8 ; 6, stamen, × 16 ; 7, ♀ flower, × 16 ; 8, older ♀ flower, × 16 ; 9, l.s. of 8, × 16 ; 10, ♀ fruiting bracteole, seen from front, × 4 ; 11, pair of ♀ fruiting bracteoles seen from side, × 4 ; 12, pair of ♀ fruiting bracteoles seen from front, with part of one bracteole removed to expose seed, × 4 ; 13, seed, × 8 ; 14, seed, l.s., showing the cotyledons above and the radicle below, both cut transversely, × 8. 1, from *Graham* 2166 ; 2–14, from *Jeffery* K380.

KENYA. Mombasa District : Mombasa, Oct. 1929, *Graham* 2166 ! ; Kilifi District : Sabaki [now Galana] R. near Malindi, Oct. 1873, *Kirk s.n.* !

DISTR. **K7** ; endemic to the coast of Kenya

HAB. In sand at or near high-tide mark

DISTR. (of species as a whole). Palestine, Sinai, Egypt, Arabia (including Aden), A.-E. Sudan, Somaliland and Kenya

Baker and Clarke's qualms about the possible misapplication of the name *A. farinosa* —see F.T.A. 6 (1) : 83 (1909)—appear unnecessary ; our plant is decidedly mealy rather than tomentose.

All authors dealing with *A. farinosa* outside our area agree in describing the fruiting bracteoles as smooth and without tubercles ; this is confirmed by the specimens. Therefore, although Forskål's type-specimen is lacking, it is reasonable to assume that the typical plant—*A. farinosa* Forsk. var. *farinosa*—was similar and to separate varietally the Kenya plants. Of the two specimens from our area cited under *A. farinosa* in F.T.A. 6 (1) : 82 (1909), that of Kirk cited above is in young fruit but shows the tubercles already visible on the bracteoles, while his Mombasa specimen is only in bud.

3. **A. semibaccata** *R.Br.*, Prodr. Fl. Nov. Holl. 1 : 406 (1810) ; Benth., Fl. Austral. 5 : 175 (1870) ; Aellen in E.J. 68 : 409, Fig. 4, C (1938) ; Aellen in E.J. 70 : 388, t. 24, C (1940) ; Black, Fl. S. Austral., ed. 2, 2 : 297, Fig. 372 (1948). Types : Australia, New South Wales, Port Jackson, *R. Brown* 2322 (BM, syn. !, K, isosyn. !, number of latter altered to 3022) ; Queensland, Keppel Bay, 1802, *R. Brown* 2322 (BM, syn. !)

Perennial herb, sometimes fruiting in its first year, or thinly woody shrub, grey to grey-green, mealy especially on leaves beneath ; stems prostrate, up to 1 m. long. Leaves oblanceolate, oblong-lanceolate or sometimes narrowly obovate, 0·5–3 (–4) cm. long, 0·2–0·8 (–1) cm. wide, rounded to acute at apex, cuneate at base, entire or with up to 6 small teeth along each margin. Flowers clustered in leaf-axils, upper clusters male and female mixed, lower female. Fruiting bracteoles free to about or below middle, when fresh their lower part swollen, smooth, fleshy, usually cherry-red, sometimes greenish, when dry flattened, veiny, and appearing cuneate into a very short pedicel ; basal connate part about 1–3 mm. long ; free apical part of each bracteole ovate-triangular, acute, 2–3 mm. long, 3–4 mm. wide, entire or with 1–5 minute teeth on each margin ; bracteoles smooth on back, rarely with 1–2 dorsal projections.

KENYA. Naivasha District, unlocalized, *Nattrass* H63/42 ! & Naivasha, 6 May 1943, *W. Andrews in Bally* 4480 ! & Lake Naivasha, 23 Mar. 1947, *Bogdan* 435 !

DISTR. **K3** ; native of Australia ; introduced into other parts of the world ; naturalized in South Africa and elsewhere

HAB. Weed of gardens and cultivated land

Bogdan 435 and *Andrews in Bally* 4480 are both var. *semibaccata* (var. *typica* Aellen), among the variants distinguished by Aellen in E.J. 68 : 409–413 (1938). Nattrass' specimen is inadequate.

4. **A. muelleri** *Benth.*, Fl. Austral. 5 : 175 (1870) ; Aellen in E.J. 68 : 369, Fig. 1, N 1–2 (1938) ; Aellen in E.J. 70 : 390, t. 24, E (1940) ; Black, Fl. S. Austral., ed. 2, 2 : 297, Fig. 374 (1948). Type : Australia, Queensland, 1869, *F. v. Mueller* (K, lecto. !)

Herb, annual or perhaps longer-living, sometimes slightly woody, ascending to erect, up to about 0·7 m. high, grey to grey-green, more or less mealy, often densely so. Leaves in outline spathulate to elliptic or rhombic-ovate, mostly 1–6 (–7) cm. long, 0·5–4 cm. wide, mostly obtuse to subtruncate at apex, gradually tapering at base, normally with 1–4 coarse teeth on each margin. Flowers clustered in leaf-axils, upper clusters male or mixed, lower female. Fruiting bracteoles normally almost sessile, their basal connate part 1–2 mm. long obconical or campanulate and becoming thick, hard and pale ; free apical part of each bracteole semicircular to triangular in outline, 2–3 mm. long, 2·5–4·5 mm. wide, acute or subacute at apex, with normally

1–4 little teeth on each margin ; bracteoles in centre smooth or venose, not at all tuberculate or muricate.

KENYA. Nakuru District : Elmenteita, Soysambu Estate, 3 Aug. 1947, *Bogdan* 1021 !
DISTR. **K3** ; native of Australia ; introduced elsewhere, and naturalized in South Africa
HAB. In colonies near stables, presumably a weed ; about 1830 m.

DOUBTFUL SPECIES
A. hastata *L.*, Sp. Pl. 1053 (1753). Recorded from **T1**, Bukoba District, in P.O.A. C : 171 (1895) and in F.D.O.-A. 2 : 204 (1932). *A. hastata* is an annual herb with triangular-hastate lower leaves truncate or shortly cuneate at base and flowers paniculate or spicate, native of Europe, N. Africa, Asia and N. America. It might occur as an introduced weed, but in the absence of any confirmatory specimens, I consider the evidence insufficient for the inclusion of this rather critical species in our flora.

3. ARTHROCNEMUM
Moq., Chenop. Monogr. Enum. 111 (1840)

Perennial herbs or thinly woody shrubs, erect or decumbent, glabrous, seemingly leafless, apparently built up of numerous, superposed, more or less tubular segments which are green and succulent and ultimately shrivel and fall away from the stem ; each segment at apex forming a little cup, usually with two short teeth, embracing the base of the next higher segment. Fertile segments aggregated into spikes at ends of stem and lateral branches, latter often very short ; spikes ultimately disarticulating or not. Flowers minute, hermaphrodite or unisexual, normally in threes, free or nearly so, a pair of threes to each fertile segment, the threes on opposite sides and immersed. Calyx irregularly toothed or lobed and obtusely pyramidal at apex. Stamens 1, rarely 2, per flower. Fruits with membranous or horny pericarp. Seeds " vertical " (i.e. laterally compressed) ; testa membranous or hard, glabrous, papillose or smooth. Embryo along one side of the seed, slightly curved or shaped like a comma, the radicle pointing downwards. Endosperm well-developed, lateral.

The segments mentioned above are morphologically interpreted as being each derived from a pair of opposite leaves, whose laminae are not developed, and whose bases are connate into a succulent sheath investing the stem ; they are decussate. The fertile segments similarly represent paired bracts, each with a sessile axillary triad of flowers.

A. indicum (*Willd.*) *Moq.*, Chenop. Monogr. Enum. 113 (1840), *pro parte* ; Ungern-Sternberg in Att. Congr. Int. Bot. Firenze 274 (1876) ; Hook. f., Fl. Brit. India 5 : 12 (1886) ; Bak. & C. B. Cl. in F.T.A. 6 (1) : 86 (1909), *quoad descr., excl. specim. cit.* Type : India, Madras Presidency, Tranquebar, *Klein in Herb. Willdenow* (B, holo., K, photo. !)

Main stems prostrate, up to at least 40 cm. long, reaching several feet and forming loose open mats in age, with profuse, lateral, ascending or erect, fertile or sterile branches about 15–30 cm. high. Sterile segments (see above) mostly about 5–11 mm. long and 3–6 mm. in diameter, when mature their sides often convex in outline. Male flowers unknown. Female flowering spikes mostly about 1–4 cm. long and 4–5 mm. in diameter, the flowers (except stigmas) hidden. Fruiting spikes cylindrical or somewhat thickened in middle, obtuse at apex, about 5–7 mm. in diameter, built up of numerous closely imbricate ring-like thickened fertile segments, their margins 1·5–3 (–4) mm. apart vertically, completely hiding the fruits ; spikes finally disarticulating. Fruiting calyces spongy, thickened, more or less obtriangular, obliquely truncate at apex, about 3 mm. long and 2 mm. wide and deep, adherent to the horny pericarp. Seed about 1 mm. long and 0·8 mm. wide ; testa pale brown, thin and rather soft. Fig. 5.

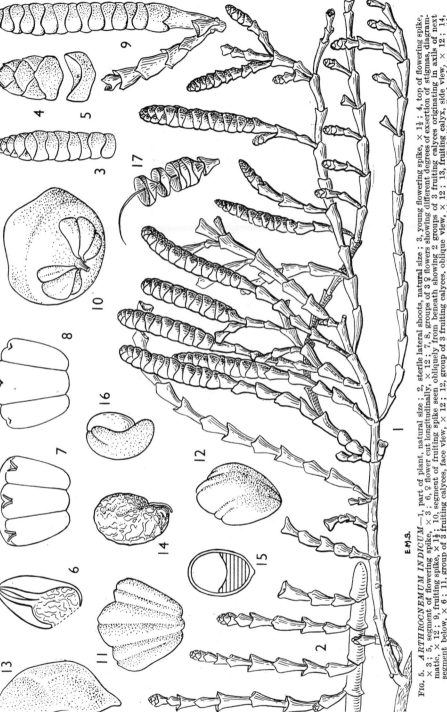

FIG. 5. *ARTHROCNEMUM INDICUM*.—1, part of plant, natural size ; 2, sterile lateral shoots, natural size ; 3, young flowering spike, × 1½ ; 4, top of flowering spike, × 3 ; 5, segment of flowering spike, × 3 ; 6, ♀ flower cut longitudinally, × 12 ; 7, 8, groups of 3 ♀ flowers showing different degrees of exsertion of stigmas, diagrammatic, × 12 ; 9, fruiting spike, × 1½ ; 10, segment of fruiting spike seen obliquely from beneath showing 2 groups of 3 fruiting calyces originating in axils of next segment below, × 6 ; 11, group of 3 fruiting calyces, face view, × 12 ; 12, group of 3 fruiting calyces, oblique view, × 12 ; 13, fruiting calyx, side view, × 12 ; 14, seed, × 24 ; 15, seed, diagrammatic, t.s. ; 16, seed with testa removed, showing embryo to left and endosperm to right, × 24 ; 17, fruiting spike breaking up, × 1½.
1, 3–6, from *MacNaughtan* 40 ; 2, 9–14, 16, from *Greenway* 4777 ; 17, from *Moss* 7072.

E.M.S.

KENYA.   Mombasa District : Port Tudor, Oct. 1931, *MacNaughtan* 40 *in Forestry Dept.* 2726 *and in C.M.* 13332!
TANGANYIKA.   Tanga District : Tanga Bay, 25 Nov. 1936, *Greenway* 4777! ; Mafia Is., Mchangani, 20 Sept. 1937, *Greenway* 5278! ; Lindi District : Kitulu Hill, 12 Feb. 1903, *Busse* 2446!
ZANZIBAR.   Zanzibar Is., Mbweni, 4 Feb. 1929, *Greenway* 1336!
DISTR.   **K**7 ; **T**3, 6 ; **Z** ; Somalia, Portuguese East Africa and coast of India from Bombay to Bengal, including Ceylon ; Angola and Senegal records very doubtful ; wrongly recorded, through confusion with allied species, from the East Indies and Australia
HAB.   Saltbush on the drier landward sides of mangrove forest, covered by high seasonal tides, often associated with the grass *Sporobolus virginicus*

SYN.   *Salicornia indica* Willd. in Ges. Naturf. Fr. Neue Schr. 2 : 111, t. 4, Fig. 2 (1799) ; Wight, Ic. Pl. Ind. Or. 3 : t. 737 (1843)

Careful search should be made for the male or hermaphrodite plant of this species. The specimens at Kew are all female, and staminate flowers eluded Roxburgh, Wight, Ungern-Sternberg, J. D. Hooker and Gamble. Moquin's statement that the flowers are " subpolyandrous " is incorrect :   Roemer & Schultes, Syst. Veg., ed. 15, 1 : 40 (1817), misquoted as " Flores 5–7 andri " Robert Brown's description of the flowers of *Salicornia indica* as " floribus 5–7 diandris " (Prodr. Fl. Nov. Holl. 1 :   411 :   1810). R. Brown's plant was, however, not an *Arthrocnemum* at all but *Salicornia australis* Soland. ex Benth. ; while Roemer & Schultes' misquotation was uncritically copied by Moquin.   Backer, in Fl. Malesiana, ser. 1, 4 (2) :   105 (1949) describes male flowers under *A. indicum* ; unfortunately the plant he is dealing with is not the same species as *A. indicum*.   The only botanist who may have seen the stamens of the real *A. indicum* is Cooke, who (Fl. Bombay, 2 : 504 : 1906) says " Anthers large, oblong-ovoid, 1/30– 1/25 in. long," without further comment. This statement stands alone, and requires confirmation.

DOUBTFUL SPECIES

" **A. fruticosum** *Moq.*"   Recorded by A. Peter, F.D.O.-A. 2 :   205 (1936) from **T**3, Tanga District, Sigi R. delta region, and from **T**6, between Dar es Salaam and Bagamoyo.   The specimens backing these records are now destroyed.

A. fruticosum Moq. is a mixture of various elements, of which the only two " possibles " from our point of view are *Salicornia fruticosa* L., not so far known from our area, and *Arthrocnemum glaucum* (Del.) Ungern-Sternb., which occurs as far south as the Anglo-Egyptian Sudan and Eritrea.   Collectors in the areas of Tanganyika mentioned above are asked to find out what Peter's plant could have been, if it has not been exterminated by the many changes that have happened along the coast.

4. **SALICORNIA**

L., Sp. Pl., ed. 5, 4 (1754)

Annual or perennial herbs, or sometimes small shrubs, glabrous, seemingly leafless, apparently built up of numerous, superposed, more or less tubular segments which are green to reddish and succulent, and ultimately shrivel ; each segment at apex forming a little cup, usually with two short teeth, embracing the base of the next higher segment.   Fertile segments aggregated into spikes at ends of stem and lateral branches, latter often very short ; spikes not disarticulating.   Flowers minute, hermaphrodite, usually in threes, more or less connate, a pair of threes to each fertile segment, the threes on opposite sides and immersed.   Calyx normally flattened at apex into the shape of a kite (i.e. cuneate with $\pm$ rounded top), in whose middle is a 3–4-denticulate opening through which stamens and stigmas project. Stamens mostly 2, sometimes 1, per flower.   Fruits with pericarp soft below, hardened in upper part.   Seeds " vertical " (i.e. laterally compressed) ;

testa thinly coriaceous, minutely hairy. Embryo filling the whole seed, shaped like a closed inverted U, the cotyledons large, and the radicle pointing downwards. Endosperm none or almost absent.

The segments mentioned above are morphologically the same as those of *Arthrocnemum* (see p. 18), to which *Salicornia* is very similar in general appearance.

S. pachystachya [*Bunge ex*] *Ungern-Sternb.* Vers. Syst. Salicorn. 51 (1866) ; in Atti Congr. Int. Bot. Firenze 304, Fig. 18 (p. 278) (1876). Type : Madagascar, *Boivin in Herb. Bunge* (? LE, holo.)

Annual herb, about 10–23 cm. high ; main stem erect with numerous suberect lateral branches. Fertile spikes many, 1·2–3·2 cm. long (to 5 cm. long, *fide* Ungern-Sternberg), 4·5–5·5 mm. in diameter, thick, cylindrical or somewhat tapering above, each of about 8–15 segments whose margins are about 1·75–2·5 mm. apart vertically, not or only very slightly hiding the fruiting calyces. Stamens 2 per flower. Central fruiting calyx of each group of three with a flattened apex about 1·5–2 mm. long and wide, when dry appearing wider than long, when softened by boiling about as long as wide ; upper margin of the apex somewhat arcuate, lateral margins straight or slightly concave. Seed ellipsoid, about 1·5–1·7 mm. long and 1–1·2 mm. wide ; testa brown. Each group of three fruiting calyces finally falling away, leaving three little cavities in the spike arranged more or less in a triangle.

TANGANYIKA. Mafia Is., Ras Mbisi, 1 Oct. 1937, *Greenway* 5361 !
DISTR. T6 ; Madagascar
HAB. Saline-water swamp forest with *Avicennia marina* and *Lumnitzera racemosa* mangroves ; growing in more open parts with *Suaeda* on a yellow sandy clay ; sea level

SYN. *Arthrocnemum pachystachyum* ([Bunge ex] Ungern-Sternb.) A. Chev. in Rev. Bot. Appl. 2 : 748 (1922), excl. saltem specim. cit. Pervillé 661

The identity of this plant is deduced solely from Ungern-Sternberg's description and figure, with which our plant fits well, except for having rather shorter spikes. Another specimen, also from Mafia Island, W. coast, 7 Aug. 1932, *Schlieben* 2646 (K !), looks very different from *Greenway* 5361, but may possibly be no more than a very young condition of the same species ; it was too shrivelled and distorted for examining the flowers.

It is impossible to reach any final decision about the taxonomy of the East African *Salicornia*, with only two gatherings to work on. More material, with branchlets preserved in spirit, and careful field-observations of habit and colour are badly wanted. Then only can its relationship be cleared up with *S. perrieri* A. Chev. from Madagascar, which may be synonymous, and with *S. brachiata* Roxb. from India, whose fruiting spikes seem more slender than in *S. pachystachya*, and which may well be distinct.

5. SUAEDA

[Forsk. ex] Scop.,* Introd. Hist. Nat. 333 (1777), excl. verba
falsa : " Capsula quinquevalvis . . . polysperma "

Annual or perennial herbs or shrubs, usually glabrous. Leaves usually fleshy, entire, usually alternate, narrow, cylindrical or somewhat flattened. Flowers axillary or sometimes adnate to the subtending leaf, solitary or clustered, subtended by small scarious bracteoles, hermaphrodite or unisexual. Calyx 5-lobed or -partite, sometimes thickened or inflated in fruit, but not winged. Stamens 5, absent or reduced to staminodes in the ♀ flowers. Ovary sessile on a broad base, or partly adnate to the calyx. Stigmas 2–5. Fruits with membranous or sometimes spongily thickened indehiscent pericarp. Seeds " horizontal " or " vertical " (i.e. vertically or laterally compressed) ; testa normally smooth, thin and hard. Embryo spiral. Endosperm absent or scanty.

* *Suaeda* is usually attributed to Forsk., Fl. Aegypt.-Arab. 69 (1775), and is even conserved from there, but Forskål provided no generic description.

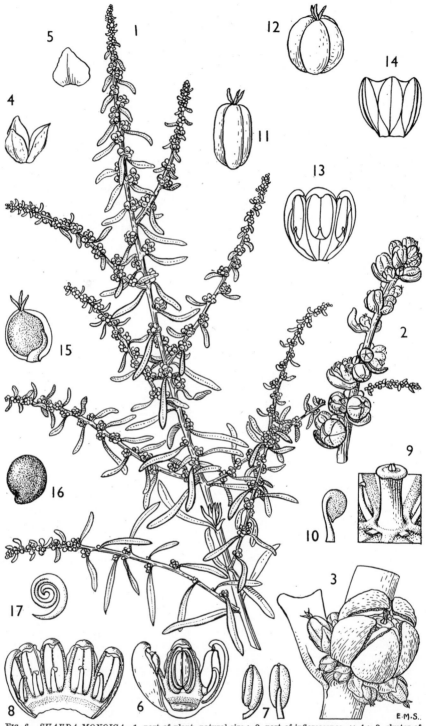

FIG. 6. *SUAEDA MONOICA*—1, part of plant, natural size ; 2, part of inflorescence, × 4 ; 3, cluster of flowers, showing larger ♂ flower surrounded by smaller ♀ flowers, × 10 ; 4, bracteoles, × 10 ; 5, single bracteole, × 10 ; 6, ♂ flower, l.s., rudimentary ovary removed, × 10 ; 7, stamens, two views, × 10 ; 8, calyx of ♂ flower opened out, seen from inside, anthers removed, × 10 ; 9, rudimentary ovary of ♂ flower in position, × 15 ; 10, rudimentary ovule of ♂ flower, × 15 ; 11 and 12, ♀ flowers, × 10 ; 13, portion of calyx of 11 seen from inside, showing staminodes, × 10 ; 14, portion of calyx of 12 seen from inside, no staminodes, × 10 ; 15, ovary with portion of membranous wall removed to show ovule in situ, × 10 ; 16, seed, × 10 ; 17, embryo, × 10. 1, 4, 5, 13–17, from *Greenway* 5271 ; 2, 3, 6–12, from *Greenway* 1337.

S. monoica [*Forsk. ex*] *J. F. Gmel.*, Syst. Nat. ed. 13, 2 (1) : 503 (1791) ; Boiss., Fl. Or. 4 : 940 (1879) ; Bak. & C. B. Cl. in F.T.A. 6 (1) : 92 (1909). Types : Egypt, Alexandria & Arabia, Lohaja, *Forskål* (C, syn.)

Shrub 1·2–6 m. high, much branched, glabrous, or occasionally with a little inconspicuous and rapidly disappearing pubescence on the very youngest parts only. Leaves fleshy, alternate, linear to linear-oblong, obtuse to acute at apex, narrowed near the sessile or very shortly petioled base, lower and median leaves mostly 1·3–3·3 (–4) cm. long, 1·5–3 mm. wide, the upper leaves (bracts) progressively shorter. Flowers green, clustered in upper axils, the clusters aggregated to form interrupted rarely dense spikes bracteate throughout or below only ; plants either ♂ ♀ (upper and median axils ♂ or ♂ and ♀, lower ♀) or ♀ only. ♂ flowers with sepals free above, 5 stamens, and a rudimentary ovary expanded at apex into a peltate disc ; stigmas not apparent or very small. ♀ flowers with sepals connate to apex, staminodes minute or none, an ovary free from calyx except at very base, and 3–4 stigmas. Fruits with membranous pericarp. Seeds " vertical " (see generic description), about 1·5–1·75 mm. long and 1–1·25 mm. wide ; testa black, glossy, almost smooth. Fig. 6.

KENYA. Masai District : Kemana, 25 June 1941, *Edwards* 249 ! ; Mombasa District : Changamwe, 31 June 1934, *Napier in C.M.* 6407 ! & Port Tudor, *MacNaughtan* 87 *in Forestry Dept.* 2649 !
TANGANYIKA. Lushoto District : Mkomazi, 23 Apr. 1934, *Greenway* 3955 ! ; Tanga, Feb. 1893, *Holst* 2111 ! ; Mafia Is., Chole Is., 19 Sept. 1937, *Greenway* 5271 !
ZANZIBAR. Zanzibar Is., Mbweni, 4 Feb. 1929, *Greenway* 1337 ! & Cooper's, 8 Apr. 1952, *R. O. Williams* 158 !
DISTR. K1, 2, ? 4 (a record from Lugard's Falls perhaps refers to this province), 6, 7 ; T2, 3, 6, 8 ; Z ; Portuguese East Africa, extending northwards to Egypt, Palestine and Syria, and eastwards to Arabia, Ceylon and India (Madras)
HAB. Saltbush on the landward sides of mangrove forest, common in places, able to resist frequent flooding by salt water, and sometimes associated with *Avicennia*. Saltbush also occurs inland on saline soils or near salt-lakes in semi-arid country ; here *S. monoica* is usually an exclusive dominant, forming thicket, open bushland or open scrub, and may be associated with the grass *Sporobolus robustus*. *S. monoica* is said to emit a peculiar smell, very perceptible and characteristic where it is an exclusive dominant ; 0–1280 m.

SYN. *S. monoica* Forsk., Fl. Aegypt.-Arab. 70 (1775), *nom. illegit.*

Schweinfurth (in Bull. Herb. Boiss. 4, app. 2 : 158–9 : 1896) describes a dwarf glaucous variant from near Aden, and other forms with vividly flesh-red coloured leaves. Similar variants or ecotypes should be looked for in East Africa.

6. SALSOLA
L., Gen. Pl., ed. 5, 104 (1754)

Annual or more often perennial herbs or shrubs, usually more or less pubescent. Leaves usually more or less fleshy, entire, alternate or sometimes opposite, sessile, narrow, sometimes scale-like. Flowers axillary, solitary or fascicled, subtended by relatively large bracteoles, hermaphrodite. Calyx (4-) 5-partite, the lobes almost always each developing in fruit a scarious horizontally spreading wing above the middle. Stamens 5, sometimes fewer. Stigmas 2, very rarely 3. Fruit with membranous or somewhat fleshy pericarp. Seeds normally " horizontal " (i.e. vertically compressed) ; testa membranous. Embryo spiral. Endosperm absent.

S. dendroïdes *Pall.*, Illustr. 22, t. 14 (1803) ; Iljin in Fl. U.R.S.S. 6 : 255 (1936). Type : Russia, *Pallas* (holo. where ? Not found at BM.)

Shrub up to about 1·5 m. high, profusely and densely branched, the branchlets erecto-patent, pubescent at least when young, whitish to pale brown ; the branchlet-systems often pyramidal in outline. Leaves very

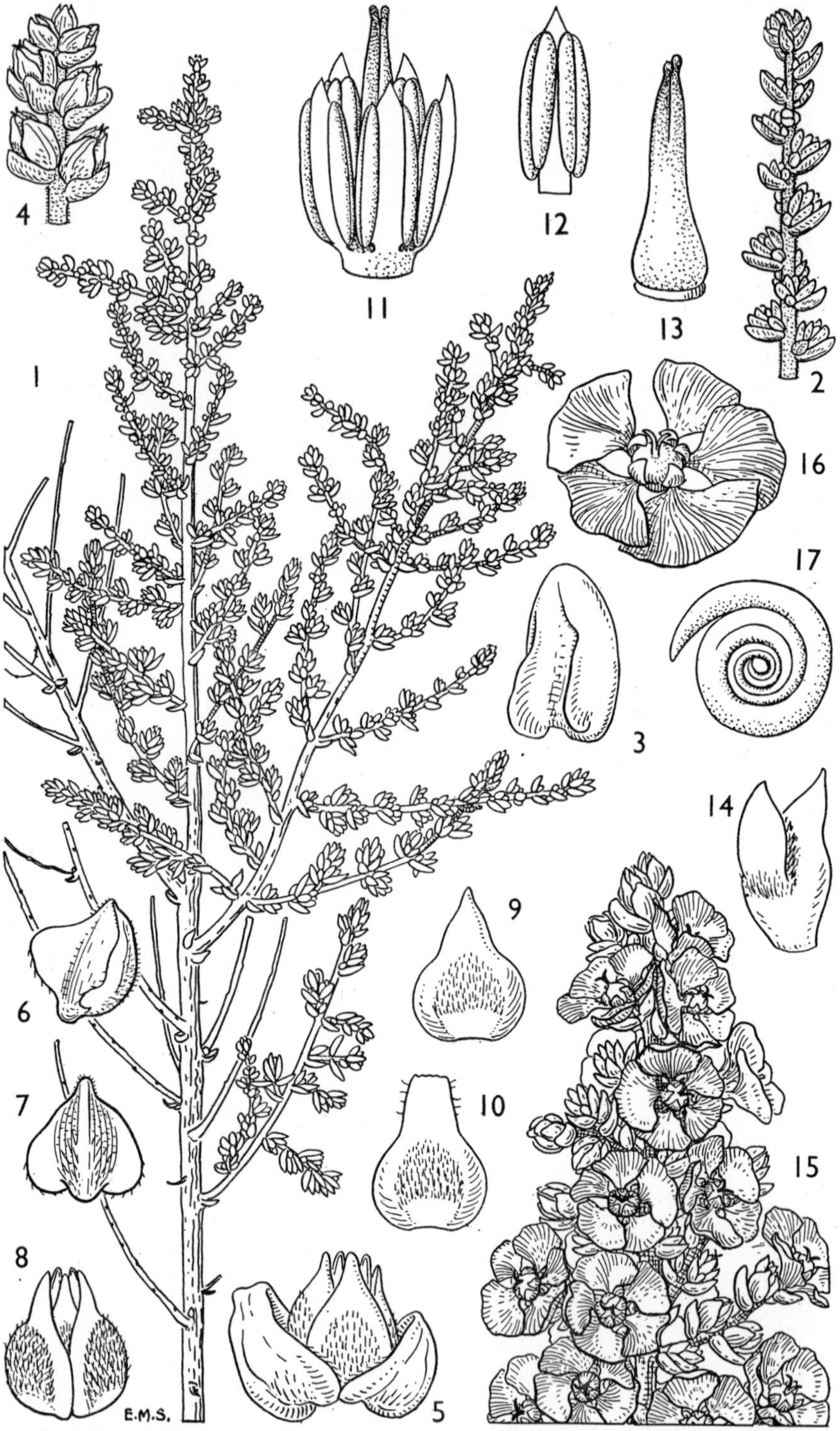

FIG. 7. *SALSOLA DENDROIDES* var. *AFRICANA*—1, part of plant, **natural** size ; 2, leafy branchlet, × 2 ; 3, leaf, × 8 ; 4, part of inflorescence, × 2 ; 5, flower with two subtending bracteoles, × 8 ; 6, bracteole, seen obliquely, × 8 ; 7, bracteole, outside view, × 8 ; 8, flower, × 8 ; 9, sepal, × 8 ; 10, sepal opened out, × 8 ; 11, stamens and styles, × 16 ; 12, stamen, × 16 ; 13, ovary, × 16 ; 14, sepal with developing wing, × 8 ; 15, mature fruits in situ, × 2 ; 16, fruit, × 4 ; 17, embryo, × 16. 1–14, from *Dalton* 7; 15–17, from *Bally* 2011.

small, fleshy, very shortly linear to boat-shaped or scale-like, about 1–4 (–8) mm. long and about 1–1·5 mm. wide, glabrous or pubescent ; the upper leaves appearing in dense tiny clusters owing to the presence of very abbreviated axillary shoots. Flowers solitary in upper axils, about 2–2·75 mm. long, sessile, often aggregated, and seeming spicate. Sepals 5, glabrous to pubescent. Fruit (including wings) about 6–10 mm. in diameter ; wings encircling the fruit, each wing reniform to transversely elliptic or deltoid-obovate, 2–4 mm. long, 1·5–7 mm. wide, scarious with a sheeny surface, very closely veined, erose on outer edge. Seed 1·5–1·7 mm. in diameter, brown.

SYN. *S. verrucosa* M. Bieb. in Mém. Soc. Nat. Moscou 1 : 103 (1812 ?) ; Boiss., Fl.
Or. 4 : 961 (1879). Type : Russia, Caspian–Caucasian plains, *Bieberstein* (? LE, holo.)

var. **africana** *Brenan* in K.B. 1953 : 433 (1953). Type : Kenya, Northern Frontier Province, Garissa, *Bally* 2011 *in C.M.* 12465 (K, holo. ! EA, iso. !)

Leaves (at least on sterile shoots) sometimes opposite, on flowering shoots shortly oblong or boat-shaped when seen from side, mostly 1–3·5 mm. long, 1–1·5 mm. thick vertically, bluntly keeled on back ; leaf-clusters comparatively loose. Fig. 7.

KENYA. Northern Frontier Province : Garissa, 28 Dec. 1942, *Bally* 2011 ! ; Turkana District, June 1934, *Buxton* 1023 ! ; Tana River District, *Battiscombe* 223 !
DISTR. K1, 7 ; not found outside our area
HAB. Semi-desert scrub and in sand near rivers in very low rainfall regions ; 90–910 m.

Asiatic material of *S. dendroïdes* seems to differ solely by minor vegetative differences : the leaves on the flowering shoots are all or mostly scale-like, about 1–2 mm. long and 1–1·5 mm. wide, the clusters of leaves more tightly bunched than in var. *africana* ; on the lower branches the foliage is similar to that of var. *africana*, except that the alternate arrangement seems constant.

DISTR. (of species as a whole). Russia from Siberia (as far E. as Balkhash and the Syr Daria foothills) to the Caspian Sea and the Caucasus, Turkish Armenia and Kurdistan, Persia, and in East Africa ; the Arabian record is an error

INDEX TO CHENOPODIACEAE